# What's Your Point of View On the Environment?

**GLOBE FEARON Educational Publisher**
A Division of Simon & Schuster
Upper Saddle River, New Jersey

**Executive Editor:** Joan Carrafiello
**Project Editor:** Renée E. Beach
**Product Development:** Pencil Point Studio
**Production Editor:** Rosann Bar
**Art Direction:** Joan Jacobus
**Interior Design:** Paradigm Design, Inc.
**Marketing Managers:** Margaret Rakus, Donna Frasco
**Electronic Page Production:** Pencil Point Studio
**Cover:** Bill Negron/Brenda Piccard

**Photo credits:** Page 3 © Courtesy of Kids F.A.C.E.
Page 21 © The Stock Market/Gabe Palmer, 1995
Page 41 © SuperStock

Printed in the United States of America
2 3 4 5 6 7 8 9 10  99

ISBN: 0-835-93334-2
C12

**GLOBE FEARON Educational Publisher**
A Division of Simon & Schuster
Upper Saddle River, New Jersey

# Contents

# The Cleanup Kids!

## Words to know

| | |
|---|---|
| **abandoned** | given up/not used any more |
| **adopt** | to take on/accept |
| **environment** | everything around us |
| **habitat** | where a plant or animal lives |
| **migrate** | to live in one place part of the year and in another for the rest of the year |
| **pollution** | harmful things in water, air, or earth |
| **recycling** | reusing trash |
| **wildlife** | animals or plants living in nature |

Some people believe that kids only think about having friends and playing, but they're wrong!  All across the United States, young people are thinking about the **environment** and making it a better place to live.  They are **recycling** paper, plastic, and glass.  They're cleaning things up!

Young people want to end **pollution**.  They are cleaning up their neighborhoods, and helping others at the same time. In Indiana, an eleven-year-old girl didn't like the litter that was piling up near the train tracks.  She found out that the woman who used to pick up the trash had died.  So the girl got a group of friends together and started a cleanup club. They cleaned up the mess around the railroad tracks. Then, they cleaned up a town park.  Now, the park in which they play is clean and pretty.  In Detroit, students raked up garbage and turned empty lots into fun playgrounds.

Kids are planting trees and gardens to make their communities more beautiful.  In Tennessee, young people planted trees in a Nashville park.  In New York City, several high school students decided to get together and work to bring an **abandoned** garden back to life.  The garden was near a home for senior citizens, who appreciated what the students did.

In 1994, $50,000 worth of pennies collected for the Children's Earth Fund was used to plant trees in countries where the forests are being destroyed.

Young people are also protecting **wildlife**. In Indiana, kids raised money to **adopt** a timber wolf. The money they raised helped protect the wolf's natural **habitat**. In 1995, teenagers again collected pennies for the Children's Earth Fund. This time, the money was used to protect wildlife that **migrate,** such as birds and sea turtles.

Kids everywhere are writing letters to their mayors, Congress, and to the president. They want the government to require people to clean up. They also are asking the government to protect animals from illegal hunting, to save the forests for natural habitats, and to stop pollution.

But many people in government say they can't force people to clean up. They also think that it is too costly to spend money on protecting animals. More money should be spent on people instead. A lot of kids disagree. We all want to grow up in a clean world, filled with tall trees, clean air, and beautiful animals. We all have that right, don't we?

## Think about this...

• How can we get everyone to join the cleanup?

• What can you do to join the cleanup?

• What will our environment be like if it is not cleaned up?

## What's your point of view?

# Checking Your Understanding

Answer these questions after reading the story.
Circle the best answer.

1. "The Cleanup Kids!" is about

    **a.** caring for animals in their habitat.

    **b.** making the environment a better place to live.

    **c.** how to plant trees.

2. According to the story, kids everywhere are

    **a.** writing letters to people in our government.

    **b.** adopting timber wolves.

    **c.** watching birds and turtles that migrate.

3. Young people are making their communities more beautiful by

    **a.** sending trees to other countries.

    **b.** planting trees and gardens.

    **c.** not caring about anything.

4. According to the story, the main reason why many kids want to join the cleanup is that

    **a.** the government is telling them to clean up.

    **b.** they could collect a lot of pennies.

    **c.** they want to live in a clean environment.

# Choosing an Issue

Choose one of the following issues.
Put a check in the box next to the issue you choose.

☐ Should the government require people to clean up?

☐ Should the government spend more money on animals and the environment?

☐ Do we have a right to grow up and live in a clean America?

✓ **CHECKPOINT**

Did you...

☑ understand what you read?

☑ choose an issue?

☐ get the facts?

☐ organize the information?

☐ take a closer look?

☐ think other ways?

☐ tell it your way?

☐ decide what is important?

☐ form your point of view?

# Getting the Facts

In order to come up with your own point of view, you need to get the facts. Get the facts by **looking**, **listening**, **remembering the story**, and by **asking questions**.

**LOOK**  Today, most communities have *recycling centers*. A recycling center is a place where trash is collected, sorted, and sent to companies that reuse it.

Look at the sign below. What information does the sign give you? Answer the questions that follow.

**1.** What does the recycling center accept? Write two things below.

_____

_____

**2.** What does the recycling center *not* accept?

_____

**3.** Which day is the recycling center open the longest?

_____

**4.** What must be done to newspapers?

_____

**5.** Which day is the recycling center closed?

_____

**LISTEN TO OTHERS**   Talk about recycling with the class. What things do you and your family recycle? Does everyone recycle the same kind of things? Make a list on the chalkboard.

How do you and your family recycle? Do you keep things to be recycled all together at home, or do you sort them? Do you bring it all to a recycle center, or does it all get picked up at home?

Is recycling important? Share your ideas.

**REMEMBER THE FACTS**   Look at the items below.
Write **T** on the line if the sentence is **TRUE**.
Write **F** if it is **FALSE**. If it is **false**, rewrite the sentence to make it **true**.
The first one is done for you.

**1.** Kids only think about having friends and playing.    _____F_____

_Young people are thinking about the environment and making it_

_a better place to live._

_____

**2.** Young people want to end pollution.    _____

_____

_____

_____

**3.** In 1994, $500 worth of pennies was collected for the Children's Earth Fund.    _____

_____

_____

_____

**4.** The government wants to spend more money on animals.    _____

_____

_____

_____

**ASK QUESTIONS**   The chart below lists things that can be recycled. Write what you already know about each item. What else do you want to know?

Fill in the chart. One is done for you. When you are finished, find the answer to your questions at the library.

| Things that can be recycled | I already know that... | What do you want to know? |
|---|---|---|
| plastic | A lot of food is packaged in plastic. | Can all plastic be recycled? |
| glass | | |
| newspapers | | |
| tin cans | | |
| car tires | | |
| metal | | |

# Organizing the Information

You can organize information by reading a cycle map.
Using a cycle map helps you:

• put your thoughts together.

• see how events relate to each other.

• think more clearly about a topic.

• show how things change.

• increase your knowledge.

• understand what happens in a logical order.

Look at the cycle map below.  Organize the information.
Write **1**, **2**, **3**, **4**, and **5** in the correct box.  The first one is
done for you.

THE WATER CYCLE

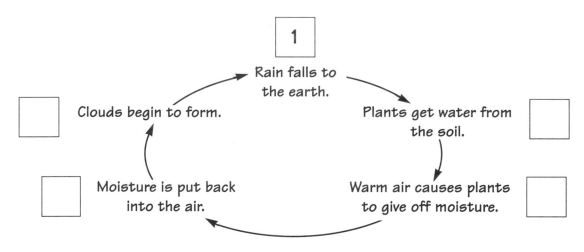

Now, answer these questions.

**1.** What is the main topic of this cycle map?

_____

**2.** What happens to the rain first?

_____

**3.** What causes plants to give off moisture?

_____

**4.** What finally happens?

_____

A **symbol** is a design that expresses an idea. Look at the symbol below. This symbol might be found on things that have been recycled.

**1.** How is the recycle symbol like the cycle map on page 8?

_____

_____

_____

_____

**2.** Why do you think the recycle symbol was designed this way?

_____

_____

_____

_____

**3.** What idea does the recycle symbol express?

_____

_____

_____

_____

_____

_____

_____

**✓ CHECKPOINT**

Did you...

☑ understand what you read?

☑ choose an issue?

☑ get the facts?

☑ organize the information?

☐ take a closer look?

☐ think other ways?

☐ tell it your way?

☐ decide what is important?

☐ form your point of view?

**4.** Look for recycle symbols around your home and in school. In your notebook, make a list of the things you find that have a recycle symbol on them. Can they be recycled again? Share your list with the class. Talk about some reasons why we recycle things.

# Taking a Closer Look

A **fact** is something that is known for sure. An **opinion** is based on what a person thinks.

**FACT:** The United States recycles 100 million tons of paper, metal, and glass each year.

**OPINION:** Some people think it is not easy to recycle.

Read each statement below. Decide if it is a **FACT** or an **OPINION**. Put an **X** on the correct line. The first one is done for you.

|  | | FACT | OPINION |
|---|---|---|---|
| **1.** | Kids only think about having friends and playing. | _____ | _____X_____ |
| **2.** | A girl from Indiana started a cleanup club. | _____ | _____ |
| **3.** | Kids are planting trees and gardens. | _____ | _____ |
| **4.** | Money was collected for the Children's Earth Fund. | _____ | _____ |
| **5.** | The government should require people to clean up. | _____ | _____ |
| **6.** | It is too costly to spend money on animals. | _____ | _____ |
| **7.** | More money should be spent on people. | _____ | _____ |
| **8.** | We all have a right to grow up in a clean world. | _____ | _____ |

Finish each statement below to make it a **fact**. Write your answers on the lines. Go back to the story if you need some help. The first one is done for you.

**1.** In Indiana, an eleven-year-old girl started a cleanup club because…

_she didn't like the litter that was piling up._ _____

**2.** Kids are planting trees and gardens because…

_____

**3.** The government does not want to spend money on animals because…

_____

**4.** We all want to grow up in a clean world because…

_____

**10** The Cleanup Kids!

Read each statement below.  They are **opinions**.
Write **PRO** if the opinion given is **for** cleaning things up.
Write **CON** if the opinion given is **against** cleaning up.  The
first one is done for you.

**1.** The government must require people to clean up.                              *PRO*

**2.** The government must protect animals from being hunted
illegally, save the forests for natural habitats, and stop pollution.    _____

**3.** People should not be forced to join the cleanup.                          _____

**4.** More money should be spent on people, not animals.                     _____

**5.** Planting trees and gardens will make our communities
more beautiful.                                                                              _____

**6.** Pollution must be stopped.                                                        _____


Why is it a good idea for the government to require all
people to join the cleanup?  Write about it in the space
below.  Use some **PROS** to support your opinion.

_____

_____

_____

_____

_____

_____


Why is it <u>not</u> a good idea for the government to require
all people to join the cleanup? Write about it in the space
below.  Use some **CONS** to support your opinion.

_____

_____

_____

_____

_____

_____

Look at each pair of events below. One sentence in each pair is a **cause**. The other is the **effect**.
A **cause** is what happens first. Write **1** by the **CAUSE**.
An **effect** is what happens last. Write **2** by the **EFFECT**.
The first one is done for you.

_2_    A girl and her friends cleaned up around the railroad tracks.

_1_    Litter was piling up near the train tracks.

_____    A garden was planted near a home for senior citizens.

_____    The senior citizens appreciated what the young people had done.

_____    Kids in Indiana adopted a timber wolf.

_____    In Indiana, kids raised money.

_____    $50,000 worth of pennies was collected for the Children's Earth Fund.

_____    Money was used to plant trees in countries where the forests are being destroyed.

_____    Money was used to protect animals that migrate.

_____    Pennies were collected for the Children's Earth Fund.

> **✓ CHECKPOINT**
>
> Did you...
>
> ☑ understand what you read?
> ☑ choose an issue?
> ☑ get the facts?
> ☑ organize the information?
> ☑ take a closer look?
> ☐ think other ways?
> ☐ tell it your way?
> ☐ decide what is important?
> ☐ form your point of view?

Write the number of the word from Column 1 which best describes the word or phrase in Column 2. One is done for you.

| Column 1 | Column 2 |
|---|---|
| **1.** abandoned | _2_   birds and sea turtles |
| **2.** wildlife | ____   forests |
| **3.** habitat | ____   an area that is no longer used |
| **4.** environment | ____   trash |
| **5.** pollution | ____   the earth, air, and water |

# Thinking Other Ways

To come up with a good point of view, you'll need to think about a situation in many different ways. You can learn how to sharpen your thinking skills by asking "What if?" Then, think about what would happen.

Think about these questions. Write your answers. Use your imagination.

1. What if people threw their trash anywhere they wanted? What would happen?

_____

_____

_____

2. What if nothing were recycled? What would happen?

_____

_____

_____

3. What if the government required all people to join the cleanup? What would happen?

_____

_____

_____

4. What if animals could talk? What would one of them tell us about pollution in the environment?

_____

_____

_____

5. What if you wanted to do more to save the environment? What would be one thing you could do?

_____

_____

_____

6. On a separate piece of paper, write your own "what would happen?" question from the story. Share your question and answer with the class.

You can learn how to sharpen your thinking skills by using information from a graph. Look at the graph below. Then, answer the questions.

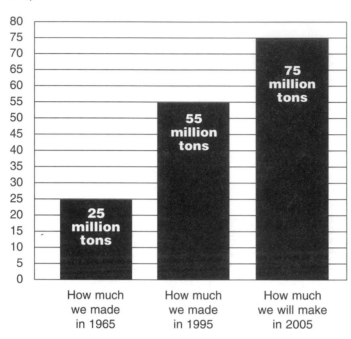

**GARBAGE IN AMERICA**

TONS
(in millions)

How much we made in 1965: 25 million tons
How much we made in 1995: 55 million tons
How much we will make in 2005: 75 million tons

**1.** In what year did Americans make the least amount of garbage? _____

**2.** How many tons of garbage will Americans make in the year 2005? _____

**3.** What has happened to the amount of garbage Americans made between 1965 and 1995? _____

## Talk about...

Look at the graph again. What is happening to all of the garbage we are making? What do you think will happen to our environment if we don't start cutting down now?

- Tell the facts.

- Talk about ways we can improve the situation.

- Make a list.

# Telling It Your Way

Look at the pictures. They are out of order. Put the pictures in the correct order. Write **1**, **2**, **3**, and **4** in the box under each picture to tell the story. Write about the story in your own words. One is done for you.

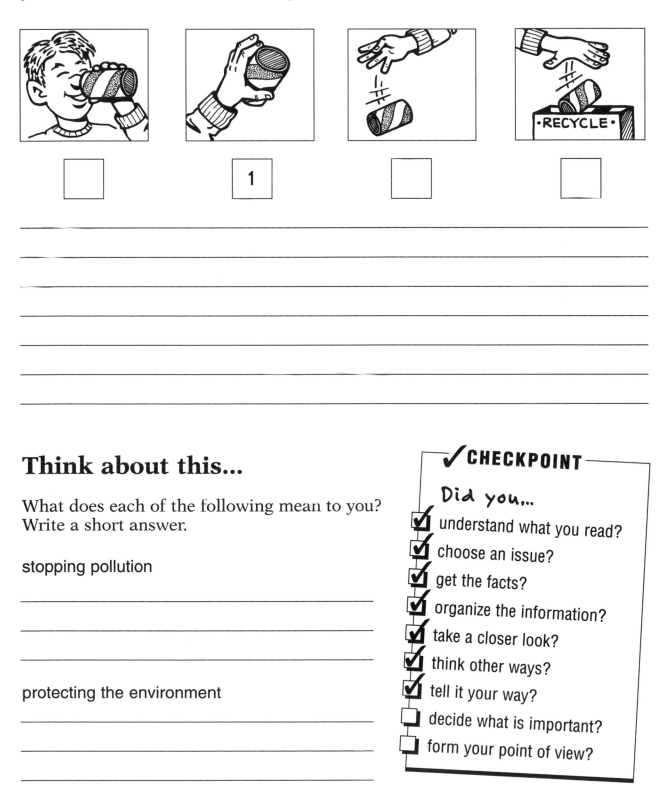

_____

_____

_____

_____

_____

_____

## Think about this...

What does each of the following mean to you? Write a short answer.

stopping pollution

_____

_____

protecting the environment

_____

_____

✓**CHECKPOINT**

Did you...
- ☑ understand what you read?
- ☑ choose an issue?
- ☑ get the facts?
- ☑ organize the information?
- ☑ take a closer look?
- ☑ think other ways?
- ☑ tell it your way?
- ☐ decide what is important?
- ☐ form your point of view?

# Deciding What Is Important

Answer the questions below. Look at your answers from the previous pages. Think about the information you need to form your own point of view.

**1.** What issue did you choose? (See page 4.)

_____

_____

**2.** Decide what is the **most important** fact you found about the story. (Look at page 10.)

_____

_____

**3.** Make a list of the pros and cons. (See page 11.)
Are they fair and reasonable?
If you agree, write **A** on the line.
If you do not agree, write **D**.

Agree? (Write **A**)

Disagree? (Write **D**)

**PROS**

_____  _____

_____  _____

_____  _____

_____  _____

_____  _____

**CONS**

_____  _____

_____  _____

_____  _____

_____  _____

_____  _____

Choose the question below that matches the issue you selected on page 4. Answer the question.

**1.** Should the government require people to clean up?
Write **yes** or **no**. _____
List some facts that support your decision.

_____

_____

_____

_____

_____

**2.** Should the government spend more money on animals and the environment?
Write **yes** or **no**. _____
List some facts that support your decision.

_____

_____

_____

_____

**3.** Do we have a right to grow up and live in a clean America?
Write **yes** or **no**. _____
List some facts that support your decision.

_____

_____

_____

_____

_____

# What's your point of view?

Present **your** point of view. Write your opinion below. Tell your readers what you think of the whole issue and why. Be sure to answer the questions you first thought about on page 3. Base your opinion on the facts and information you have found. Share your work with the class.

the questions you first thought about on page 3.

✓ **CHECKPOINT**

Did you...
- ☑ understand what you read?
- ☑ choose an issue?
- ☑ get the facts?
- ☑ organize the information?
- ☑ take a closer look?
- ☑ think other ways?
- ☑ tell it your way?
- ☑ decide what is important?
- ☑ form your point of view?

# Going a little further...

## On Your Own

**1.** The *Environmental Protection Agency* in Washington, D.C. is responsible for making sure we all live in a clean environment. They offer free information to everyone on pollution, protecting animals, and recycling. Write a letter to the *Environmental Protection Agency*. Ask for information. Look at what they send you and write a report. You can also draw pictures. Present your report to the class.

*For information about protecting the environment, write to:*

Environmental Protection Agency
Public Information Center, M/C 3404
401 M Street, SW
Washington, D.C. 20460
(202) 260-2080

## Work with a Partner

**2.** Earth Day is a day we celebrate cleaning up the environment. Every April 22, people in America talk about issues such as pollution, trash, and protecting wildlife.

Work with a partner. Together, decide what **Earth Day** means to you both. Decide to do one special thing together on Earth Day to save the environment. Tell the class about what you and your partner plan to do.

**3.** Together, design your own Earth Day symbol. You can use trees, birds, animals, people or anything else you want to draw in your symbol design. Write **Earth Day, April 22, (and the year)** at the bottom of your design. Display your Earth Day symbol in class. Talk about your symbol and why you and your partner decided to draw it this way.

## Work in a Group

**4.** Work as a class. You will need a very large cardboard box and a bathroom scale. Put the cardboard box in the corner of the room. Write **RECYCLE** on the box. Use the box to trash paper only. Each day, throw all of your paper trash in the box. Weigh the box at the end of each week. How much does the box weigh in one week? in one month? Talk about what you learned with the class. Recycle the paper properly when you are finished.

**5.** Start a cleanup campaign. Together, look for litter around your school. Collect, sort, and recycle it properly.

## Words to know

| | |
|---|---|
| **algae** | a simple plant-like life form that grows in water |
| **fertilizers** | chemicals or other things that are added to soil to make plants grow |
| **nonpoint pollution** | pollution that comes from many sources |
| **point pollution** | pollution that comes from one source |
| **runoff** | rainwater flowing over the land |
| **sewage** | wastes from our homes and from farm animals |

Before 1972, the condition of our nation's rivers and lakes was terrible. Many of them were polluted with factory wastes and **sewage**. Rivers were turning red from dyes from factories. Fish and plant life in the waters died. In 1972, the United States passed the Clean Water Act, which made polluting the waters illegal, and gave money to towns and cities to clean up their lakes and rivers. The waters became cleaner.

The Clean Water Act helped people fight **point pollution**. Point pollution is pollution that comes from one source, like waste from a factory. A harder kind of pollution to control is **nonpoint pollution.** Nonpoint pollution is polluted **runoff** that comes from lots of sources, like the dirty water that runs down your street after a heavy rain. Nonpoint pollution contains chemicals like bug sprays, paints, motor oil, gasoline, **fertilizers**, and bathroom waste.

Fertilizers and animal wastes from nonpoint pollution can cause **algae** to grow in lakes. Algae needs oxygen to live. It robs oxygen from the water, which causes fish and shrimp to die. In some places, fertilizers in the water have made algae grow so fast that the lakes and rivers are clogged.

Stopping nonpoint pollution is hard because it comes from so many different places. Imagine trying to control the runoff from your own backyard!

At present, more than half of the nation's water pollution is caused by farming. People are asking farmers to plant crops in different areas each year, to plow less, and to stop using bug sprays. Some farmers think this is unfair. They are worried that they will have smaller crops and make less money.

Cities and towns are also trying to reduce nonpoint pollution. They are thinking about changing building codes, which are the rules builders have to follow when they make a house. Builders may have to add special pipes to hold runoff water. This would be expensive for them. Then, it would probably cost more for people to buy a home.

Many students are trying to help the pollution problem by joining special programs such as Save Our Streams. Save Our Streams (SOS) asks students to adopt a section of a nearby stream. Every month or two, the students go to the stream to collect and test samples of the water. They count the number of life forms in the water and send the results to a national office of SOS, which tells if the stream is healthy or becoming polluted. What can you do?

## Think about this...

• How can we stop point and nonpoint pollution?

• What can you do to be sure our water is clean?

• How clean is your water?

## What's your point of view?

# Checking Your Understanding

Answer these questions after reading the story.
Circle the best answer.

1. "What's in the Water?" is about

    a. farmers who plant crops.
    b. pollution in our rivers and streams.
    c. motor oil, gasoline, and bathroom wastes.

2. According to the story, the Clean Water Act

    a. stopped runoff.
    b. helped pollution come from one source.
    c. made polluting the waters illegal.

3. Fertilizers and animal wastes in our waters promote

    a. the growth of algae.
    b. the Clean Water Act.
    c. the oxygen fish need to live.

4. Many students are helping pollution problems by

    a. counting the number of plants in the water.
    b. adopting a section of a nearby stream and
    testing the water.
    c. visiting the national office of Save Our Streams
    once a month.

# Choosing an Issue

Choose one of the following issues.
Put a check in the box next to the issue
you choose.

☐ Is it the responsibility of the government
alone to be sure that our waters are clean?

☐ Should everyone be required to take
responsibility for keeping our water clean?

☐ Are tougher laws needed to control both
the use of fertilizers and the removal of
our wastes?

✓CHECKPOINT

Did you...

☑ understand what you read?
☑ choose an issue?
☐ get the facts?
☐ organize the information?
☐ take a closer look?
☐ think other ways?
☐ tell it your way?
☐ decide what is important?
☐ form your point of view?

## Getting the Facts

In order to come up with your own point of view, you need to get the facts. Get the facts by **looking**, **listening**, **remembering the story**, and by **asking questions**.

**LOOK**   Learn how to use a dictionary. Look up the meaning of each word shown below. Write the meaning in your own words on the lines.

| freshwater | saltwater | stagnant |
| rainwater | purify | |

**freshwater** _____

_____

_____

**saltwater** _____

_____

_____

**stagnant** _____

_____

_____

**rainwater** _____

_____

_____

**purify** _____

_____

_____

**LISTEN TO OTHERS**  Talk about the kind of animals that live nearby or in a river or stream.  How would polluted water affect them?  Would there be any affect on the trees and plants that grow around the river or stream?  How would it affect people who live along the river or stream?  Listen to your classmates' ideas.

**REMEMBER THE FACTS**  Look at the items below.  Write **T** on the line if the sentence is **TRUE**.  Write **F** if it is **FALSE**. If it is **false**, rewrite the sentence to make it **true**.  The first one is done for you.

1. Before 1972, rivers and lakes were clean.                 _____F_____

   _Before 1972, rivers and lakes were in terrible condition._

   _____

   _____

2. The Clean Water Act helped people fight nonpoint pollution.    _____

   _____

   _____

   _____

3. In some places, fertilizers in water made algae grow so
   fast that the lakes and rivers became clogged.            _____

   _____

   _____

   _____

4. Stopping pollution is easy because it comes from only
   one place.                                                _____

   _____

   _____

   _____

**ASK QUESTIONS**  The best way to gather information is by asking questions. First choose a topic. Then, ask yourself, **what do I want to know?** Write some questions. Finally, ask **how will I find out?** Make a list of some of the ways you can find out.

**Topic:**  *Fish that live in the ocean.*

**What do I want to know?**

*How do fish breathe?*

*What do fish eat?*

*How do fish swim?*

*What is the biggest danger to fish in the ocean?*

**How will I find out?**

*Go to the library.*

*Read a science book about fish and oceans.*

*Look in an encyclopedia.*

Choose your own topic about water pollution. Then, ask yourself, **what do I want to know?** Write some questions. Finally, ask **how will I find out?** Make a list of some of the ways you can find out.

**Topic:** _____

**What do I want to know?**

_____

_____

_____

_____

**How will I find out?**

_____

_____

_____

# Organizing the Information

You can organize information by creating a flow chart.
A flow chart helps you:

• put your thoughts together.

• see how facts relate to each other.

• think more clearly.

• increase your knowledge.

• understand topics.

Look at the words below.  Use them to fill in
the blanks on the flow chart.  Write the correct
word in the box where it belongs.  Some are
done for you.

**fertilizers      factories      bathrooms      wastes**

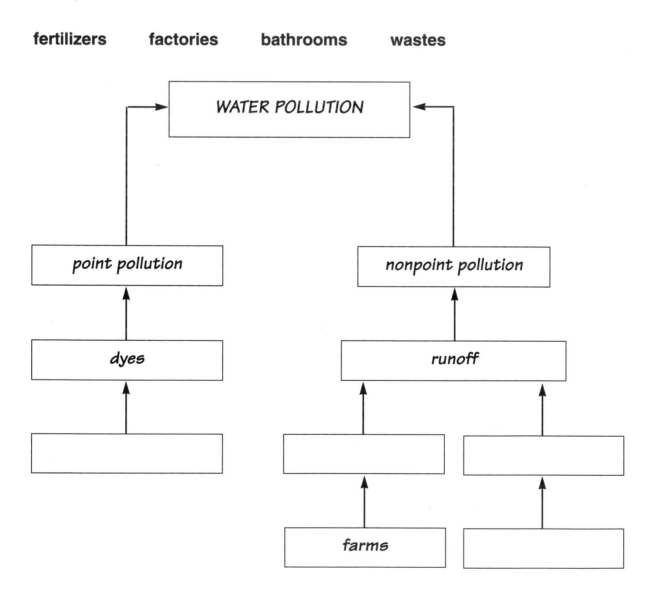

Try another one.  Look at the words below.
Use them to fill in the blanks on the flow
chart. Write the correct word in the box
where it belongs.

**runoff      streams      rivers      oceans**

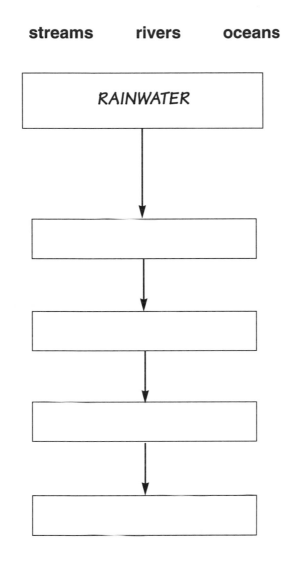

**1.** What is the topic of this flow chart?

_____

**2.** Why do you think this is called a flow chart?

_____

_____

_____

**3.** On a separate piece of paper, make up your own flow chart.
What topic will you use?  What words will you need to
describe the topic? Put the words in the correct order on
your flow chart.

Look at the information from the story. Write the letter of the phrase in the correct column below. The first one is done for you.

A. try to control point pollution
B. ask farmers to plant crops in different areas each year
C. allow farmers to use fertilizers and bug sprays so they can have a bigger crop and make more money
D. keep building codes the same so that builders do not have to add special pipes
E. follow the rules that were passed in 1972 under the Clean Water Act
F. do not try to control nonpoint pollution because it comes from so many places
G. give more money to the towns and cities to clean up their lakes and streams
H. do not think about nonpoint pollution because the runoff will take everything away

## ✓ CHECKPOINT

Did you...

☑ understand what you read?
☑ choose an issue?
☑ get the facts?
☑ organize the information?
☐ take a closer look?
☐ think other ways?
☐ tell it your way?
☐ decide what is important?
☐ form your point of view?

**Things that would stop water pollution**

_____A_____

_____

_____

_____

**Things that would cause water pollution**

_____

_____

_____

_____

Now, look at all of the phrases again. Which do you think is the <u>one most important thing</u> to do that would stop water pollution? Write it in the space below.

_____

_____

_____

Why did you choose this phrase?

_____

_____

_____

# Taking a Closer Look

A **fact** is something that is known for sure. An **opinion** is based on what a person thinks.

**FACT:**       Plants and animals need water to live.

**OPINION:**    The water we drink does not taste good.

Read each statement below. Decide if it is a **FACT** or an **OPINION**. Put an **X** on the correct line. The first one has been done for you.

|  | FACT | OPINION |
|---|---|---|
| **1.** Before 1972, rivers were turning red from dyes. | X | |
| **2.** The Clean Water Act made polluting illegal. | | |
| **3.** Algae robs oxygen from the water and the fish and shrimp die. | | |
| **4.** It is unfair to ask farmers to plant crops in different places each year. | | |
| **5.** People think farmers should not use bug sprays. | | |
| **6.** Adding special pipes in homes would probably make homes cost more to buy. | | |
| **7.** Stopping nonpoint pollution is hard. | | |
| **8.** Many students are joining special programs to take care of pollution problems. | | |

Write the number of the word from Column 1 which best describes the word or phrase in Column 2. One is done for you.

| Column 1 | Column 2 |
|---|---|
| 1. point pollution | _____ dirty water after the rain |
| 2. helps crops grow | _____ nonpoint pollution |
| 3. runoff | _1_ dyes from a factory |
| 4. algae | _____ fertilizers |
| 5. sewage | _____ leaves lakes clogged |

Think about this opinion.

**The government must require people to reduce pollution.**

Now, read each sentence below. Write **PRO** if the sentence gives a good reason why people **should** be required to reduce pollution. Write **CON** if the sentence gives a reason why people **should not** be required to reduce pollution. The first one is done for you.

1. Before the Clean Water Act, our rivers and lakes were in poor condition.                    _*PRO*_

2. Nonpoint pollution is too difficult to control because it comes from so many sources.          _____

3. Too much fertilizer and animal wastes in lakes cause algae, and algae kills fish.             _____

4. Some farmers do not care how much bug spray they use because they want to have more crops and money.   _____

5. Forcing builders to follow tough building codes would make the price of buying a home too costly.     _____

6. Some students are finding that the stream they adopted is polluted.                            _____


Why do you think it is a good idea for the government to require people to reduce pollution? Write about it in the space below. Use some **PROS** to support your opinion.

_____

_____

_____

_____


Why do you think it is difficult for the government to require people to reduce pollution? Write about it in the space below. Use some **CONS** to support your opinion.

_____

_____

_____

_____

Take a closer look at water pollution. Do the following experiment on your own, with a partner, or as a class. You will need the following items.

**3 clear jars**      **paper towels**      **a spoonful of vegetable oil**

## Here's what to do:

1. Fill the first jar with water from the tap.
2. Fill the second jar with water from a puddle or rainwater runoff from around your school or home.
3. Fill the third jar with water from the tap. Add a spoonful of vegetable oil to the water.
4. Follow the directions to experiment with each jar.

### JAR 1: TAP WATER

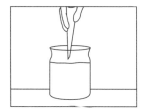

a. Roll one paper towel into a cone shape. Dip it into the first jar and mix around.
b. Take the paper out of the jar and look at it. What did you see happen? Write a short sentence below.

### JAR 2: PUDDLE WATER

a. Rolling another paper towel into a cone shape, dip it into the second jar and mix it around.
b. Take the paper out of the jar and look at it. What did you see happen? Write a short sentence below.

### JAR 3: TAP WATER AND VEGETABLE OIL

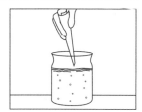

a. Rolling a third paper towel into a cone shape, dip it into the third jar and mix it around.
b. Take the paper out of the jar and look at it. What did you see happen? Write a short sentence below.

c. Now, try to clean the oil off the water with another rolled paper towel. Is it easy or hard to do? Explain your answer.

Please clean up when you're done and don't drink the water!

# Thinking Other Ways

To come up with a good point of view, you need to think about a situation in many different ways. You can learn how to look at an issue in other ways by asking "What if?"

Think about the experiment on page 31. Then, look at these questions. Write what you think would happen. Use your imagination.

1. What if Jar 3 were the ocean and oil spilled into it? What would happen?

_____

_____

_____

_____

_____

_____

_____

_____

_____

_____

_____

_____

_____

2. What if Jar 2 were the ocean? What would happen to the plants, fish, and shrimp that live in the ocean? Why?

_____

_____

_____

_____

_____

_____

_____

You learned about water pollution and what people are trying to do to clean things up. But how much water do we waste? Look at the facts below. Then, answer the questions.

---

### WHAT WE CAN DO TO SAVE WATER

**1.** Don't run the water while you brush your teeth.

**2.** Take short showers instead of long baths.

**3.** Use water-saving shower heads and faucets.

**4.** Save a flush.

**5.** Collect rainwater and use it for watering plants.

**6.** Fix faucets that leak.

**7.** Do larger loads of laundry, not one thing at a time.

**8.** Wash dishes quickly. A dishwasher also saves water.

**9.** Keep a bottle of water in the refrigerator. Then you will have cool drinking water.

Remember, everything that goes down the drain ends up in a river or ocean, or in someone else's water!

---

**1.** What is one thing that you can do to save water that you are not already doing?

_____

_____

_____

_____

**2.** What is one thing that your family can do that they are not already doing at home?

_____

_____

_____

_____

**3.** Why is it important to save water?

_____

_____

_____

_____

How much water do you use?  Look at the facts below.
Then, answer the questions.

---

### HOW MUCH WATER USED EACH TIME

| | | |
|---|---|---|
| **1.** | Brushing your teeth with water running (3 minutes) | 12 gallons |
| **2.** | Short shower (3 minutes) | 9 gallons |
| **3.** | Flushing (per flush) | 6 gallons |
| **4.** | Bath (per bath) | 50 gallons |
| **5.** | Laundry (per load) | 30 gallons |
| **6.** | Washing dishes (per load) | 10 gallons |

Each person uses about 30,000 gallons of water each year!

---

**1.** What one thing uses the most water?

_____

_____


**2.** Compare the amount of water used to brush your teeth with the amount used to take a short shower.  What happens?

_____

_____


**3.** Do you think you use 30,000 gallons of water each year? Why or why not?

_____

_____


# Talk about...

Talk about the foods you eat and drink every day. Make a list.  Does each food on your list contain water?  Are they cooked with water? What would happen if our water was polluted?

- Tell the facts.

- Tell the cause and effect.

- Talk about what else could happen.

## Telling It Your Way

Write a short paragraph to tell what the story is about.
Be sure to use some of the new words you have learned.
Look at page 20 if you need some help.

_____

_____

_____

_____

_____

_____

_____

_____

_____

_____

_____

_____

## Think about this...

What does each of the following mean to you?
Write a short answer.  Look them up if you
need some help.

environmental protection

_____

_____

toxic waste

_____

_____

✓ **CHECKPOINT**

Did you...

☑ understand what you read?

☑ choose an issue?

☑ get the facts?

☑ organize the information?

☑ take a closer look?

☑ think other ways?

☑ tell it your way?

☐ decide what is important?

☐ form your point of view?

# Deciding What Is Important ▬▬▬▬

Answer the questions below.  Look at your answers from the previous pages.  Think about the information you need to form your own point of view.

**1.** What issue did you choose?  (See page 22.)

_____

_____

**2.** Decide what is the **most important** fact you found about the story.  (See page 29.)

_____

_____

**3.** Make a list of the pros and cons.  (See page 30.)
Are they fair and reasonable?
If you agree, write **A** on the line.
If you do not agree, write **D**.

Agree? (Write **A**)

Disagree? (Write **D**)

**PROS**

_____    _____

_____    _____

_____    _____

_____    _____

_____    _____

_____    _____

**CONS**

_____    _____

_____    _____

_____    _____

_____    _____

_____    _____

_____    _____

Choose the question below that matches the issue you selected on page 22. Answer the question.

1.  Is it the responsibility of the government alone to be sure that our waters are clean?
    Write **yes** or **no**. _____
    List some facts that support your decision.

    _____
    _____
    _____
    _____
    _____

2.  Should everyone be required to take responsibility for keeping our water clean?
    Write **yes** or **no**. _____
    List some facts that support your decision.

    _____
    _____
    _____
    _____
    _____

3.  Are tougher laws needed to control both the use of fertilizers and the removal of our wastes?
    Write **yes** or **no**. _____
    List some facts that support your decision.

    _____
    _____
    _____
    _____
    _____

# What's your point of view?

Present **your** point of view. Write your opinion below. Tell your readers what you think of the whole issue and why. Be sure to answer the questions you first thought about on page 21. Base your opinion on the facts and information you have found. Share your work with the class.

the questions you first thought about on page 21.

✓ **CHECKPOINT**

**Did you...**
- ✓ understand what you read?
- ✓ choose an issue?
- ✓ get the facts?
- ✓ organize the information?
- ✓ take a closer look?
- ✓ think other ways?
- ✓ tell it your way?
- ✓ decide what is important?
- ✓ form your point of view?

# Going a little further...

## On Your Own

**1.** Draw a picture of the United States. Show the bodies of water around it. Look at a map if you need some help. What oceans are on each side? Write the names of these oceans on your map.

The body of water below the United States is called a **gulf.** A gulf is a large body of water that is part of the ocean, but has land around it. Write the name of the gulf that is to the south of the United States on your map.

Find the five lakes that border Canada, our neighbor to the north. What are these lakes called? Write the name of these lakes on your map.

Can you find the Mississippi River? Label it on your map.

When you are finished, display your map with the rest of the class. What do you notice about all of these bodies of water? Talk about how these bodies of water are all connected. What would happen if the lakes to the north were polluted?

## Work with a Partner

**2.** Go back and look at page 34. Do you and your partner each use 30,000 gallons of water each year? Work together with your partner to figure it out. Use a calculator. Choose some things from the chart that you do. For example, if you brush your teeth every day with the water running, in 7 days (1 week) you would have used 84 gallons of water!

**(12 gallons of water x 7 days = 84 gallons)**

Remember, there are 52 weeks in a year.
Who uses more water each year? you or your partner?

## Work in a Group

**3.** How can you help other people learn how to save water? As a group, plan a good campaign. Make posters, banners, and advertisements. Display them around your classroom or in the halls. You can also design "SAVE THE WATER" buttons to give to friends and family. Use different colors of paper, crayons, and paint. Then, put double-sided tape on the back to wear the buttons.

## Words to know

| | |
|---|---|
| **executives** | people in charge at a company |
| **imperative** | necessary/very important |
| **incarcerate** | to put someone in jail because of a crime he or she committed |
| **liable** | being responsible for |
| **tainted** | having a small amount of something bad in it |
| **vials** | small bottle made of glass |
| **workplace** | a place where people go to work |

Many people feel that company **executives** should be held **liable** for the environmental crimes of their companies. This means, for example, that if a company pollutes a river by allowing dangerous chemicals to get in the water, the leaders of the company should be punished.

In New York, a group of eighth-grade students found a bunch of broken blood **vials** on a beach. When the blood was tested, it was found to be **tainted** with a dangerous disease.  Police found out where the vials came from. A man who owned a blood-testing company had the vials dumped into the water. He was charged for this crime.

In Mississippi, the owner of a small oil company felt he followed all the rules for protecting the environment.  When one of the machines belonging to his company broke down and spilled ten barrels of oil, his company quickly cleaned up the spill.  Yet, the owner was still heavily fined and told that he committed a crime by not being careful.  He felt this was not fair.

Questions were asked at some large companies. Only half of the executives interviewed feel they are responsible for environmental crimes.  Some say that many times, what happens is just an accident.  Or, they say they did not know what happened and they feel they should not be blamed.

Many judges don't want to **incarcerate** people any more. Instead of being sent to jail, people are being sent to join environmental groups. They have to attend meetings and learn about how to protect the land and water, and how to stop pollution. Many people think this is a good idea.

Some people, however, do not agree. There is the case of two hunters who went hunting in an area that was not open to hunters. As punishment, they were forced to join a club whose members did not believe in hunting. The hunters felt this was unfair. Hunting, to them, is okay if it is done in the right place.

Others think that it is **imperative** to teach students more about protecting the environment. If students are educated about the environment, they will be better prepared once they get into the **workplace**.

One thing is for sure. People everywhere have to pay more attention to caring for the environment. Or they may be charged with a crime.

## Think about this...

- What could you do to learn more about protecting the environment before you get into the workplace?

- Why do some people still break the rules and risk being punished?

- Do environmental accidents just happen or is it because people are not careful?

# What's your point of view?

# Checking Your Understanding

Answer these questions after reading the story.
Circle the best answer.

1. "It's a Crime!" is about

   a. judges who don't like putting people in jail.
   b. eighth-grade students who went to the beach.
   c. environmental crimes.

2. Many judges want people to be

   a. sent to jail for committing crimes against the environment.
   b. sent to meetings to learn about the environment.
   c. sent to school to become prepared for the workplace.

3. Putting dangerous chemicals into the water is
   an example of

   a. red dye.
   b. an environmental crime.
   c. a punishment.

4. According to the story, people everywhere must

   a. pay more attention to the environment.
   b. listen to the judges.
   c. get into the workplace.

# Choosing an Issue

Choose one of the following issues.
Put a check in the box next to the issue
you choose.

☐ Is it fair to blame an executive if he or she
did not know an accident happened?

☐ Should executives be required to attend
meetings on environmental protection?

☐ Are you learning enough about protecting
the environment so that you are prepared
when you get into the workplace?

✓ CHECKPOINT

Did you...

☑ understand what you read?
☑ choose an issue?
☐ get the facts?
☐ organize the information?
☐ take a closer look?
☐ think other ways?
☐ tell it your way?
☐ decide what is important?
☐ form your point of view?

# Getting the Facts

In order to come up with your own point of view, you need to get the facts. Get the facts by **looking**, **listening**, **remembering the story**, and by **asking questions**.

**LOOK**    Look at the **pie chart** below. It is called a pie chart because it looks like a pie cut into slices. A pie chart helps you see the whole picture and the parts within. Look at the pie chart. Then, answer the questions.

**PUBLIC OPINION WHEN ASKED:
SHOULD EXECUTIVES BE LIABLE
FOR THEIR COMPANY'S ENVIRONMENTAL CRIME?**

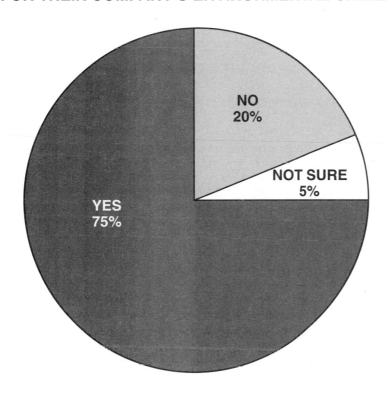

**1.** How do 20% of the people feel about this question?

_____

**2.** What percent of the people questioned were not able to give a yes or no answer?

_____

**3.** What can you tell by looking at this pie chart?

_____

_____

Talk about the environment with the class. What have you learned in school about ways to keep the environment clean?

What kind of work would you like to do when you finish school and get into the workplace? What kinds of things do you think you will have to do at your job to keep the environment clean? Share some ideas with the class.

**REMEMBER THE FACTS**   Look at the items below.
Write **T** on the line if the sentence is **TRUE**.
Write **F** if it is **FALSE**. If it is **false**, rewrite the sentence to make it **true**.
The first one is done for you.

1. Some people think that if a company pollutes, the executives should not be liable.                                          *F*

   *Some people think that if a company pollutes, the executives*

   *should be held liable.*

   _____

2. A man in New York was charged with the crime of dumping vials of tainted blood on the beach.                  _____

   _____

   _____

   _____

3. When ten barrels of oil spilled into the earth, the owner of the company was found innocent of the crime.        _____

   _____

   _____

   _____

4. Judges are sending people to jail to learn how to build new homes.                                                 _____

   _____

   _____

   _____

**5.** Twenty percent of the executives feel they are responsible for environmental crimes.

_____

_____

_____

_____

**6.** Students need to prepare for getting into the workplace.

_____

_____

_____

_____

**ASK QUESTIONS**  Learn more about a topic by asking questions. First choose a topic about the environment.  Then, ask yourself **what do I want to know?  What do I already know?  How will I find more information?** Fill in the blanks.  Some are already done for you.

**Topic:** _polluted rivers_ _____

**What do I want to know?**

_____

**What do I already know?**

_dangerous chemicals get into the water_ _____

**How will I find more information?**

_____

Choose a topic about the environment.  Write your topic on the line.  Then, answer the questions.

**Topic:** _____

**What do I want to know?**

_____

**What do I already know?**

_____

**How will I find out?**

_____

# Organizing the Information

By finding and organizing details you can:

• look at the issue closely.

• show the importance of supporting details.

• see how the issue and details relate to each other.

• put your thoughts together.

• understand the issue.

Look at each statement. Then, read the sentences that follow. Two sentences support the issue. One does not. Draw a line through the one sentence that is **not a supporting detail**. The first one is done for you.

**STATEMENT:**

Today, people are still careless about polluting.

**SUPPORTING DETAILS:**

Someone dumped vials of blood on a beach.

~~We all live in an environment.~~

Ten barrels of oil spilled when a machine broke down.

**STATEMENT:**

Environmental accidents should not be a crime.

**SUPPORTING DETAILS:**

Hunters like to hunt in the right places.

A machine broke and spilled oil into the earth.

Executives did not know an accident happened.

**STATEMENT:**

Students must learn more about the environment.

**SUPPORTING DETAILS:**

Students need to know about testing blood.

Students may be charged with an environmental crime.

Not enough is being done to educate students about pollution.

You can organize information by putting things in order.

Decide **what happened first**. Write **1** on the line.

Then, decide **what happened next**. Write **2** on the line.

Write **3** on the line to show **what happened after that**.

Then, **what finally happened**? This one is done for you.

_____ The blood was tested and found to be tainted.

_____ Eighth-grade students found a bunch of broken vials of blood.

_____ The police found out who dumped the vials.

**What finally happened?** _The owner of a company was charged for the crime._

These three events are numbered in order for you.
What finally happened? Write your answer on the line.

_1_ A machine broke.

_2_ Ten barrels of oil spilled into the earth.

_3_ The company cleaned up the oil.

**What finally happened?** _____

_____

_____

Write the number of the word from Column 1 which best describes the word or phrase in Column 2. One is done for you.

**Column 1**   **Column 2**

1. executives    _____ incarcerates people

2. judge    _____ a workplace

3. a company    _____ forced to join a club

4. the hunters    _1_ liable for environmental crimes

✓ **CHECKPOINT**

Did you...

☑ understand what you read?

☑ choose an issue?

☑ get the facts?

☑ organize the information?

☐ take a closer look?

☐ think other ways?

☐ tell it your way?

☐ decide what is important?

☐ form your point of view?

# Taking a Closer Look

Each statement below is **false**. Read each sentence. Decide which word makes it false. Cross out the incorrect word. Write the correct word above the one you crossed out. Make the sentence **true**. Go back to the story if you need help. The first one is done for you.

1. The hunters joined a club that did not believe in ~~pollution~~ *hunting*.

2. None of the executives interviewed feel they are responsible for environmental crimes.

3. People have to go to jail to learn how to protect the land and water.

4. If a company allows chemicals to get into the water, the company leaders should be freed.

5. Judges need to be taught more about protecting the environment.

Now, choose one of the statements you made true from above. Write it below.

**STATEMENT:**

_____

_____

_____

Is it a **fact** or **opinion**? **Remember:** A fact is something that is **true**. An **opinion** is based on what a person thinks. Write fact or opinion on the line.

**FACT OR OPINION:** _____

What do you think about the statement you chose? Write your opinion below.

_____

_____

_____

Read each statement below.  Write **PRO** if the statement supports the opinion that people should be held responsible for protecting the environment.

Write **CON** if the statement supports the opinion that people should not be held responsible.  One is done for you.

1. When companies dump dangerous chemicals into a river, they are poisoning our drinking water.

<u>PRO</u>

2. Company executives cannot be blamed for accidents they are not aware of.

_____

3. Hunters feel that they should have the right to hunt anywhere they want.

_____

4. People who destroy the land are destroying an important resource.

_____

5. Children need to be taught how to protect the land and water.

_____

6. Sometimes the government has to clean up environmental accidents even though they were caused by companies.

_____

Now, go back and look at your answers.

Find the statements that support people being held responsible for environmental crimes.  Write the numbers of those statements under **PROS**.

Find the statements that support people not being held responsible for environmental crimes. Write the numbers of those statements under **CONS**.
One is done for you.

**PROS**            **CONS**

<u>   1   </u>          _____

_____          _____

_____          _____

Why do you think people should be liable for polluting the environment? Write about it in the space below. Use some **PROS** to support your opinion.

_____

_____

_____

_____

_____

Why do you think people should not be liable for polluting the environment? Write about it in the space below. Use some **CONS** to support your opinion.

_____

_____

_____

_____

_____

Why do you think it is a good idea for everyone to learn more about the environment? Write about it below. Use some **PROS** and **CONS** to support your opinion.

_____

_____

_____

_____

_____

What can be done to make our environment a cleaner place to live? Write some of your ideas below.

_____

_____

_____

_____

_____

Think about each pair of events below. One sentence in each pair is a **cause**. The other is the **effect**.
A **cause** is what happens first. Write **1** by the **CAUSE**.
An **effect** is what finally happens. Write **2** by the **EFFECT**.
The first one is done for you.

_2_   The company was given a fine.

_1_   A company allowed dangerous chemicals to get into the river.

_____   The blood in the vials was tested and found to be tainted.

_____   Students found a bunch of broken blood vials on the beach.

_____   Questions were asked at some large companies.

_____   Half of the executives feel they are responsible for environmental crimes, and half feel they are not.

_____   Two hunters were forced to join a club that they had no interest in.

_____   Two hunters hunted where it was not allowed.

_____   Students are taught more about ways to protect the environment.

_____   Students are better prepared for the workplace.

Can you think of another cause and effect?
Write them on the lines below.

**CAUSE:** _____

_____

_____

**EFFECT:** _____

_____

_____

_____

# Thinking Other Ways

To come up with a good point of view, you'll need to think about a situation in many different ways.

When you get a job, you may need to make some decisions that affect the environment. Look at each question below. Then, circle the best answer.

1. Tina works in a coffee shop. She cooks french fries. What does she do with the oil when she is finished?

   a. She throws the oil down the drain.
   b. She puts the oil into containers and recycles it.
   c. She gives it to her boss.

2. Dave runs a printing press. He wants to help save the environment. What can he do?

   a. He can wash his hands when they get full of ink.
   b. He can use ink that has oil in it.
   c. He can print on recycled paper.

3. Rita is an office assistant. She uses a lot of paper. What should she do?

   a. She should put the paper in a box for recycling.
   b. She shouldn't worry about recycling.
   c. She should ask her boss to buy her a computer.

4. Juan owns a tire shop. One of his workers threw old tires in the trash. Juan got a fine because tires must be recycled. What should he do?

   a. He should fire the worker.
   b. He should talk to the worker about how to recycle tires.
   c. He should not pay the fine because he was not the one responsible.

5. Kim doesn't like her job in the school cafeteria because she is always picking up empty soda cans. What should she do?

   a. She should quit her job.
   b. She should yell at the students.
   c. She should put a sign on a trash can that says "recycle soda cans here."

Asking "What if?" helps you explore an issue. Look at your answers on page 52. Then, read the questions. Each one will help you to think about a situation in a different way. Write your answers on the lines.

**1.** What if Tina threw the oil down the drain? What would happen to the drainpipe?

_____

_____

_____

_____

_____

_____

**2.** What if Dave used ink that had oil in it? What would happen?

_____

_____

_____

_____

_____

_____

**3.** What if Rita didn't care about recycling? What would happen?

_____

_____

_____

_____

_____

_____

_____

**4.** What if Juan talked to the worker about how tires should be recycled? What would happen?

_____

_____

_____

_____

_____

_____

_____

**5.** What if Kim put a sign on a trash can that said "recycle soda cans here." What would happen?

_____

_____

_____

_____

_____

_____

_____

# Talk about...

Talk about a time when you were blamed for something you did not do. How did you feel about it? What happened? Then, think about a time when <u>you</u> blamed someone for something he or she did not do. How did you feel about it? Try to use what you have learned.

- Choose an issue.
- Put things in order.
- Tell the cause and tell the effect.
- Talk about what happened.
- Talk about what else could have happened.
- What finally happened?

# Telling It Your Way

Write a short paragraph to tell what the story is about.
Be sure to use these words in your summary. Review the
definitions on page 40.

**executives**     **imperative**     **incarcerate**     **liable**     **workplace**

_____

_____

_____

_____

_____

_____

_____

_____

_____

_____

_____

_____

_____

# Think about this...

What kind of job would you like to have?
Write what you would like to do below.

_____

How would learning about protecting the
environment help you in your job? List
some ways below.

_____

_____

_____

_____

_____

## ✓CHECKPOINT

**Did you...**

- ☑ understand what you read?
- ☑ choose an issue?
- ☑ get the facts?
- ☑ organize the information?
- ☑ take a closer look?
- ☑ think other ways?
- ☑ tell it your way?
- ☐ decide what is important?
- ☐ form your point of view?

# Deciding What Is Important

Answer the questions below. Look at your answers from the previous pages. Think about the information you need to form your own point of view.

**1.** What issue did you choose? (See page 42.)

_____

_____

**2.** Decide what is the **most important** fact you found about the story. (See page 48.)

_____

_____

**3.** Make a list of the pros and cons. (See page 49.)
Are they fair and reasonable?
If you agree, write **A** on the line.
If you do not agree, write **D**.

Agree? (Write **A**)

Disagree? (Write **D**)

**PROS**

_____   _____

_____   _____

_____   _____

_____   _____

_____   _____

**CONS**

_____   _____

_____   _____

_____   _____

_____   _____

_____   _____

Choose the question below that matches the issue you selected on page 42. Answer the question.

1. Is it fair to blame an executive if he or she did not know that an accident happened?
   Write **yes** or **no**. _____
   List some facts that support your decision.

   _____
   _____
   _____
   _____
   _____

2. Should executives be required to attend meetings on environmental protection?
   Write **yes** or **no**. _____
   List some facts that support your decision.

   _____
   _____
   _____
   _____
   _____

3. Are you learning enough about protecting the environment so that you are prepared when you get into the workplace?
   Write **yes** or **no**. _____
   What can you do to learn more?

   _____
   _____
   _____
   _____
   _____

# What's your point of view?

Present **your** point of view. Write your opinion below. Tell your readers what you think of the whole issue and why. Be sure to answer the questions you first thought about on page 41. Base your opinion on the facts and information you have found. Share your work with the class.

answer the questions you first thought about on page 41.

## ✓CHECKPOINT

### Did you...

- ☑ understand what you read?
- ☑ choose an issue?
- ☑ get the facts?
- ☑ organize the information?
- ☑ take a closer look?
- ☑ think other ways?
- ☑ tell it your way?
- ☑ decide what is important?
- ☑ form your point of view?

_____

_____

_____

_____

_____

_____

_____

_____

_____

_____

_____

_____

_____

_____

_____

_____

_____

_____

_____

_____

_____

# ═Going a little further...═

## On Your Own

**1.** What do you do to protect the environment? Think about some things you do at home and in school. Do you recycle newspapers and soda cans? Do you sometimes pick up litter to keep the environment clean? Make a list. Then, add one more thing you can do to protect our environment.

## Work with a Partner

**2.** Together, find some information about an oil spill that happened in the past few years. Newspapers, magazines, and books from the library are good places to look.

Find out where the oil spill happened. How did it happen? Was it an accident? What was the effect of the oil spill on plants, animals, fish, and people? How was the oil spill cleaned up? Was anyone blamed for it? Make posters about the oil spill to show and discuss with the class.

**3.** Many animals are dying because of pollution. If this continues, many animals will become **extinct.** What does this word mean? Together, find out about the animals that are becoming extinct. Make a list. Then, list some of the reasons why they are becoming extinct. Brainstorm about what can be done to help. If we pollute our world and we are not careful, could we be in danger of becoming extinct? Talk about your ideas with the class.

## Work in a Group

**4.** You will need small paper cups (one for each person in your group), a marker, a package of grass seed or marigold seeds, potting soil, water, table salt, some vegetable oil, and a cup to use to water the plants.

Divide the class into three groups.

Write your name and your group number on the cup with a marker. Fill your paper cup with some potting soil. Plant the seeds. Put your cup by a window or under a grow light. Keep the cups from each group together. Water your plant each day.

Group 1: use clear tap water

Group 2: use a mixture of tap water with a spoonful of salt

Group 3: use a mixture of tap water and a spoonful of oil

Watch as the plants grow. What happened? Why?

# Putting It Together!

Now, you can form your own point of view for any issue.
Use the thinking skills you have learned.

## 1. Understand what you read
- read carefully
- read the story again if you need to
- choose the issue
- learn meanings of new words

## 2. Get the facts
- look and listen
- remember the facts
- ask questions
- find out more about the issue

## 3. Organize
- use web charts and tables
- compare and contrast
- put things in order of importance

## 4. Take a closer look
- look at details
- find facts and opinions
- decide if the information presented is accurate
- look at the pros and cons
- find the causes and effects

## 5. Think other ways
- predict and reason
- find a new meaning
- solve problems
- turn the issue around

## 6. Tell it your way
- summarize
- relate what you learn to the real world

## 7. Decide what is important
- make a decision
- tell your opinion

## What's your point of view?